Ceramic Toilets

Munroe Blair

A unique illustration of Gladstone Pottery Museum's Junction toilet, manufactured by Shanks of Barrhead, Scotland.

Henson Editorial Services
&
North Staffordshire Press Ltd

Published by North Staffordshire Press Ltd
The Business Village, Staffordshire University, 72 Leek Road, Stoke-on-Trent, Stafffordshire, ST4 2AR, United Kingdom

Copyright © 2012 Munroe Blair

A catalogue record for this book is available from the British Library
ISBN: 0-9568-1986-4

Acknowledgements:-
The accuracy of information contained in this book was achieved through the help and guidance of many people with expertise in the subject. The author is indebted to those who contributed towards ensuring the historical accuracy of toilet development. Space does not permit naming everyone who so willingly devoted their time and advice; this help has been sincerely appreciated, thank you all. The extensive help by the following individuals deserves special mention: Javeriah Abbas, Paul Mellor, Terry Woolliscroft, Twyfords former Archivist, Angela Lee, Manager, Gladstone Pottery Museum for her patience during photographic and research sessions, Roger Cooper, Ideal-Standard, David Woodcock, The National Museum of Science and Industry, Miranda Goodby, Potteries Museum and Art Gallery, Gaye Blake Roberts, Director and the dedicated Wedgwood Museum team, Luitwin Gisberg von Boch-Galhau, Villeroy and Boch.

All rights reserved
No part of this publication may be reproduced, stored in a retrieval system or transmitted, in any form or by any means without the prior permission in writing of the publisher, nor be otherwise circulated in any form of binding or cover other than that in which it is published and without a similar condition including this condition being imposed on the subsequent purchaser.

Cover and book design by Paul Mellor
Printed and bound in The United Kingdom
Typeface: Baskerville (John Baskerville, 1706-1775)

Contents

Early water closets (or toilets)	1
The first water closet patents	2
Evolution of ceramic toilets	3
Sewers and the ceramic industry	6
Pedestal toilets	8
Wash out toilets	10
Wash down toilets	13
Ceramic materials used for toilet production	14
Stoneware	14
Cane and white ware	15
Earthenware	15
Vitreous china	16
Fireclay	16
Decoration	16
Single coloured ware	22
Flushing efficiency, syphonic and close coupled toilets	23
Syphonic toilet development	25
Single trap syphonic toilets	26
Double trap syphonic toilets	27
Close coupled toilet suites	27
American bathroom manufacturers	29
Worldwide sanitaryware production	30
Development in recent years	31
Toilets of the third millennium	32
Glossary	34
Places to visit	37

1   Early water closets (or toilets)

John Harington 1561-1612

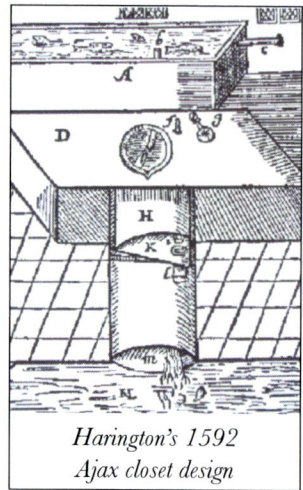
Harington's 1592 Ajax closet design

The godson of Queen Elizabeth I, Sir John Harington, designed the first water closet, the Ajax, in 1592. So impressed was Elizabeth by the Ajax that she ordered one for Richmond Palace. With the exception of a water sealed outlet trap, Harington's sixteenth century closet included most features of early twentyfirst century toilets.

The inspiration for Harington's water closet design could have been the ancient plumbing in the Roman Aquae Sulis Baths, four miles (6km) away from his Kelston home. The recorded cost of Harington's Ajax water closet was thirty shillings and eight pence. This price would have exceeded the annual income of a contemporary farm labourer.

Two centuries later, water closets remained expensive, affordable only by the wealthy. In 1664, diarist Samuel Pepys recorded having a form of closet called a "privy" discharging into a cesspit in his cellar. Pepys wrote how gangs of men known as

Roman Bath Aquae Sulis, Bath, England

"gongfermors" used buckets to empty his cesspit, carrying the stinking contents through the house. The emptying arrangements available to Pepys were retained until early Victorian times, when "night soil men" took over the "gongfermors" emptying service.

By the end of the eighteenth century, many wealthy householders installed "privies", but these insanitary arrangements invariably discharged into cesspits rather than drains. In city homes, a few cesspits lay outside the buildings, but the majority were situated beneath dwellings in domestic cellars.

Samuel Pepys 1659 -1703

Wedgwood's 1830 bourdalou

Wealthy females also had provision of toilet facilities when they travelled about the country. Horse-drawn stagecoaches took passengers over long distances, with each non-stop stage taking several hours. Ceramic receptacles called a "bourdalou," also known as "coach pots," were secreted beneath lady passengers' skirts for urinary relief on long stages. Male travellers may have had stronger bladders, but were more likely to have stood to urinate through the moving stage coach door.

The first water closet patents

Alexander Cummings a London watchmaker, registered the first valve type "Water Closet" patent in 1775. To operate Cummings' closet the handle was pulled upwards. This action simultaneously opened the outlet valve and a water cock to create a cleansing water swirl inside the bowl. Cummings' design had a "slider valve" to close the outlet hole from cesspit odours. Rust often impaired the sliding outlet plate's operation. Failure to open the valve did not deter householders from using the closet until it overflowed. Water sealed overflows were introduced to by-pass any blockages. Trapped overflow horn arrangements, as shown in the Bramah patent drawing, below right remained a feature of valve closets for nearly two hundred years.

Locksmith, engineer and inventor Joseph Bramah, adapted Cummings' idea and registered the second water closet patent in 1778. Bramah's design, with a self cleansing hinged outlet valve, formed an efficient watertight seal and soon outsold Cummings' pattern. The hinged seal was designed to hold water in the bowl and stop cesspit gas entering the dwelling it served. Introduction of Bramah's improved valve closet had taken almost two hundred years since Harington's original water closet.

Within twenty years of Bramah's patent, he claimed sales of six thousand valve type closets. Wealthy families without valve closets experienced no inconvenience, because domestic staff carried ablutionary water in and emptied away excretory effluent from chamber pots, jerrys and commodes.

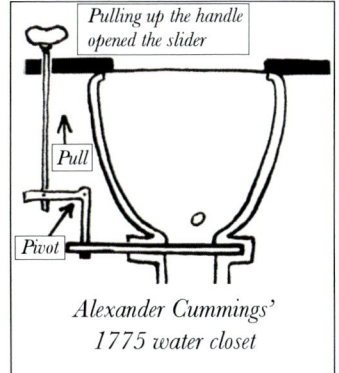
Alexander Cummings' 1775 water closet

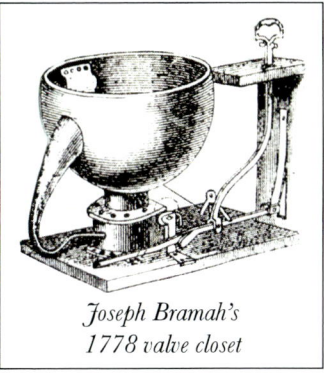
Joseph Bramah's 1778 valve closet

Bedroom chair commode with chamber pot and lid

Water closets installed in upmarket eighteenth century homes were flushed from manually operated water control valves or storage cisterns. Mains water was seldom available to cleanse these early toilets. During the eighteenth century piped mains water had been introduced in some British cities, but supplies were unreliable. Most mains water only flowed for one or two hours on two or three days a week, with pressure and flow varying. Filling roof space storage tanks to provide a constant pressure water supply required the manual opening of a mains water-cock control valve at appropriate times.

In 1748 ball valves were introduced to control water flow into roof space storage tanks, serving household needs by gravity. This service replaced the domestic staff chore of carrying water up through the dwelling for use by the family. Availability of mains water led to increased numbers of toilets, creating an intolerable demand on overloaded sewage systems.

Early metal valve closets were operated by cumbersome and noisy metal apparatus concealed in wooden cabinets. These early units suffered from constant malfunction, adding to the health hazard they were intended to prevent. Valve closet manufacturers organised component supplies and often completed their installation with wooden cabinets.

Due to the slow pace of sanitary engineering progress, the basic design of valve closets remained unaltered for a hundred years.

1809 Twyford bowl on a metal valve closet

Metal valve closet problems included the likelihood of flushing water spilling over, but more important customers were dissatisfied with the painted finish of the cast-iron bowls. Josiah Wedgwood, recorded making a ceramic "close stool water closet pan" in April 1777. Wedgwood made pans to inventor Richard Edgeworth's valve closet design drawing. Several valve closet manufacturers, including Joseph Bramah, purchased pans and bowls from Wedgwood and other Staffordshire potters.

Evolution of ceramic toilets

Seeking alternatives to the rust prone metal bowls, valve closet manufacturers worked with potters to develop pottery bowls or pans for their units. Early pottery bowls were requested with flanges for bolting onto metal components and spigot connections to the valve closet plumbing for flushing rim spreaders and overflows.

Underside of a pottery valve closet bowl with fitted metal flange, integral flushing rim and overflow spigots

Decorated pan with copper flushing spreader plate

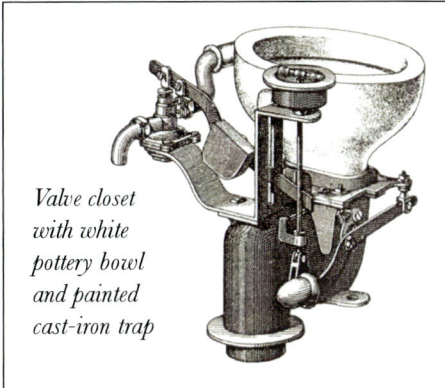

Valve closet with white pottery bowl and painted cast-iron trap

Additional features offered with early pottery pans included holes to secure copper flushing water spreader plates to help cleanse the inside of the bowl. Ceramic bowls were easier to clean, lighter in weight, not liable to corrosion and considerably cheaper than their metal predecessors. In addition, white pottery bowls could be decorated to match the wash bowls in bathrooms. The commercial opportunity to replace metal valve closet bowls with pottery alternatives, gave pioneer sanitary potters the incentive to develop an all ceramic toilet.

Potters received only a few shillings for their bowls, but householders paid very high prices for complete metal valve closets. Even though metal valve closets were very expensive, the wealthy preferred their plumbing arrangements concealed in mahogany cabinets to compliment their bathroom décor.

When flushing directly from water services, closets were considered plumbed-in or fixtures. In 1782, John Gaittait patented a stink-trap designed to close out cesspit odours. These water sealed traps, recognised today as U-bends, were adopted as a key element of toilet design. Large bore traps were expensive to fabricate in metal, but easily made in pottery. With ease of production in their favour, pottery toilets were poised to challenge the monopoly held by metal valve closets.

Potters' co-operation with valve closet makers and sanitary engineers had established the health value of water sealed self-cleansing traps to eliminate sewer odours. With the aim of securing a larger share of the emerging toilet market, it was essential for potters to by-pass the valve closet manufacturers.

Although noisy and less hygienic, valve closets enclosed in mahogany cabinets were visually more attractive to Victorian householders than pottery bowl and separate trap toilets. More hygienic pottery toilets of superior flushing performance were available, but sales remained poor. Sanitary engineers encouraged potters into a more active role in toilet design.

Decorated pan with overflow and flushing spigot

Potters had gained extensive flushing experience, by making simple pan and water sealed trap toilets for the growing cheaper industrial housing market sector. These simple toilets with separate bowls and supporting trap were known in the plumbing trade as Cottage or Liverpool pans. To challenge the luxury metal valve closet market, potters drew on their developing success with the efficient operation of pottery toilet bowls.

Cane and white Cottage pan and trap

The introduction of pottery toilets, with a separate pan and full bore glazed trap, virtually eliminated the blockage risk previously experienced with valve closets. The other important element of Cottage pan toilets was their separate water sealed traps to arrest sewer gas whilst supporting the bowls. However, the external appearance of Cottage type toilets with separate bowls and traps proved unattractive in the luxury market.

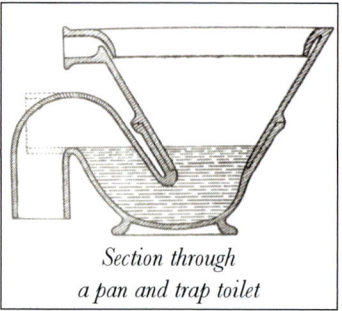

Section through a pan and trap toilet

Demand for ceramic closet bowls and traps was initially small by comparison with sales of bourdalous, chamber pots, washing bowls and water jug sets. Normally wealthy home owners could only see the white or decorated pottery bowls of their valve closets. The noisy operational metal and trap were hidden within cabinets.

Doulton mahogany encased valve closet with white bowl

The inset valve closet bowls illustrate the only ceramic portion that would be visible to the Victorian user. The right-hand picture shows the bolt holes for fitting a metal flushing spreader.

Sewers and the ceramic industry

During the eighteenth century's industrial revolution, large numbers of agricultural workers moved into urban cities and industrial areas seeking work. In these expanding conurbations drinking water was often polluted and sewage disposal systems inadequate or non-existent. The resultant increased discharge of untreated sewage rapidly overwhelmed natural rivers and waterways, reducing them to choked open sewers. In 1832, drinking water drawn from London's contaminated rivers was responsible for the first of several horrific cholera epidemics. In 1840, John Snow, who became the anaesthetist to Queen Victoria, linked cholera and typhoid fever with contaminated water supplies. Inefficient drains exacerbated the situation, as cholera and typhoid claimed the lives of 29,000 Londoners between 1832 and 1866. During the same period over 50,000 lives were lost to cholera throughout Britain.

Sir Henry Doulton Portrait by Frederick Sandys, 1861

Repeated cholera outbreaks forced the government into action. The 1848 Public Health Act stated: "it shall not be lawful newly to erect, or to rebuild any House, without a sufficient Water closet, Privy or Ashpit." This was the first Health Act to make provision for sanitary facilities inside all new dwellings.

The 1848 Public Health Act encouraged potters to develop efficient free-standing pottery toilets. In common with many cities around the world, nineteenth century London was without an effective self-cleansing sewer system.

Legislation alone could not beat cholera, but the Act started a national drive to provide efficient sewage drainage systems. Henry Doulton (1820-1897) was amongst the first to identify the commercial potential for drain pipes to connect dwellings into London's anticipated new sewer system. The projected need for drain pipes encouraged Doulton to open a stoneware drainpipe factory in Lambeth, London in 1846. For his contribution to stoneware drainpipe development, Henry Doulton was knighted by Queen Victoria in 1887. He was the first potter to receive a knighthood.

Model of a domestic dwelling house drainage system in London's Science Museum

In the 1888 *"Sanitary Record"* Sir Henry reflected on the appalling and insanitary conditions that had existed in 1850. Doulton commented that: 'pottery had then (in 1850) scarcely any use in connection with house sewerage or town drainage. It is not too much to say that the manufacture and use of pottery has advanced side by side with engineering sanitary science.'

The vital public hygiene contribution by stoneware drainpipes were the key factor in securing Doulton's knighthood; he had grasped the opportunity and helped to satisfy the demand. These ceramic stoneware drain pipes heralded the rebirth of the ancient Knossos Palace and Roman ceramic drains from several millennia earlier.

Royal warrants were granted to manufacturers who made a continuing, direct and significant supply of goods, ordered and paid for by the British royal household. Thomas William Twyford's company received recognition with a Royal Warrant of Appointment to Queen Victoria's Government, for his contribution to water closet design. King Edward VII awarded Sir Henry Doulton a Royal Warrant for the company's contribution to sanitary science.

Sir Henry was given permission in 1901 to call the company "Royal Doulton". King Edward VII granted Thomas Crapper a Royal Warrant; King George V similarly appointed Ideal Standard.

Having missed the attention of King Edward Vll, Edward Johns of Armitage incorporated a badge on their Dolphin Toilet "NOT BY APPOINTMENT TO THE KING" above a fake British Royal coat of arms.

Edward Johns' Dolphin toilet

Edward Johns' fake transfer coat of arms

Pedestal toilets

By the middle of the nineteenth century toilets ceased to be considered a luxury, instead being regarded as a necessity to conquer urban disease. Ceramic bowl and trap toilets gradually began to dominate the volume market. Many well-known potters, such as Edward Johns, John Ridgeway, Minton, Spode, Twyford, Wedgwood, Enoch Wood and Whieldon were making pottery water closet bowls, traps, plug wash bowls, portable bidets, boudalous and toiletware.

Amongst the most successful sanitary earthenware pioneers, were the Twyford brothers Christopher and Thomas. The younger brother, Thomas Twyford (1827-1872), committed his family's domestic pottery in Bath Street, Hanley, to the exclusive manufacture of sanitary pottery in 1848. Twyford was the first factory to risk industrial scale production for this specialised market with his range of water closet basins and plug outlet washbasins.

Thomas Twyford's 1848 Bath Street factory, the world's first factory devoted exclusively to sanitaryware - inset is Twyford's 1865 advertisement

London's Great Exhibition of 1851 highlighted the Victorian comparative lack of interest in free standing, plumbed in ceramic toilets. The exhibition catalogue recorded only John Ridgway of Shelton, Staffordshire displaying a ceramic water closet. Pandering to Victorian taste in bathroom décor, John Ridgway's water closet emulated in ceramic the external appearance of the popular Victorian mahogany closet cabinet. Internally, Ridgway's design simply concealed a Bramah style metal valve closet within his ceramic shell. Far more influential than Ridgway's exhibit was George Jennings public toilets introduced at the 1851 Exhibition. Jennings probably helped to start the flood tide of sanitary reform throughout Victorian Britain.

Exhibition catalogue illustration of Ridgway's 1851 water closet

During the last quarter of the nineteenth century, the majority of pottery toilet bowls and water sealed traps remained as separate items. Development work was progressing to join the bowl and trap together. Potters knew that, until they could supply toilets to satisfy Victorian aesthetic taste, their ceramic items would remain concealed in cabinets. The challenge was to create a free standing toilet by enclosing the bowl and trap inside a supporting pedestal.

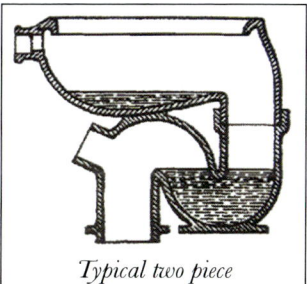

Typical two piece wash out closet

The wash out and plunger type closets were early steps towards achieving a one piece toilet. Until the complex trappage was concealed within a supporting pedestal, they would remain unsuitable for exposed installation. Health conscious authorities helped by recommending free-standing toilets, to give access for detection and repair of faulty outlet joints. Substantial potting expertise was required to create and promote free standing toilets.

To develop his 1852 patented wash out water closet (WC) design, George Jennings cooperated with Thomas Twyford, who had produced his 1851 exhibition toilets. Ceramic water sealed trap toilets had established their value by eliminating the need for expensive metal valve outlet contraptions. Unlike metal valve closets, nineteenth century toilets were designed to stand outside cabinets, but Victorians remained reluctant to accept them.

In addition to drain pipes, Henry Doulton designed and attempted to attract buyers with his free standing decorated stoneware toilet. Despite the enamelled relief decoration, there was very little interest in Doulton's stoneware two piece toilet design.

Very few established potters were prepared to follow Twyford's 1848 lead by switching their production exclusively to the heavier materials required for the much larger sanitaryware items. Most industrial size potters decided to concentrate on decorative or table top ware. Over the next sixty years Doulton, Edward Johns, Alfred Johnson, Johnson Brothers, Howsons and Shanks gradually set up plants or purchased smaller closet bowl makers to switch exclusively into sanitaryware production.

Many attempts were made to persuade Victorian buyers to accept free standing exposed pottery toilets. Efficiency, hygiene and the attractive decoration possible on pottery pedestal toilets, gradually led to the decline of metal valve closets and their expensive cabinets.

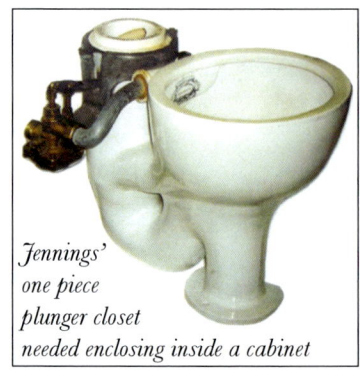

Jennings' one piece plunger closet needed enclosing inside a cabinet

Doulton's 1898 two piece free standing relief decorated toilet

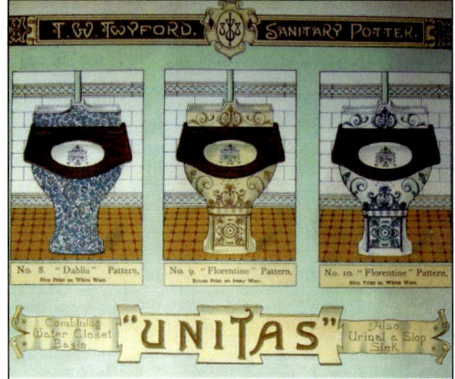

Twyford's 1888 catalogue showed the varied designs of transfer decoration and bespoke hand enamelled colours traced in gold

Britain's developing industrial sanitary potteries were mainly located in North Staffordshire. The original reasons for this were the availability of local clay, plus long flame coal to fire and salt for glazing the ware. The supportive expertise of transfer and hand painting decoration also established itself in the area. The Staffordshire Potteries acquired a worldwide reputation for quality decorative sanitary pottery and table ware.

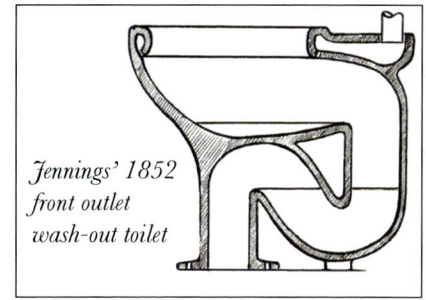

Jennings' 1852 front outlet wash-out toilet

Wash out toilets

The first step towards the replacement of valve closets had been the introduction of ceramic bowls. George Jennings, the Victorian sanitary engineer, installed early versions of his patent pending wash out water closet in the public toilets at the 1851 Crystal Palace International Exhibition. Jennings' wash out bowl featured a shallow dished tray holding an inch of water to prevent fouling inside the bowl and tray surfaces. Progressive development of Jennings' two piece pan and trap design paved the way for important improvements in free standing one piece toilet design.

George Jennings had fought hard to secure a contract for the installation of public toilets in the 1851 Great Exhibition's Crystal Palace. Records show that 827,280 (14 per cent) of visitors paid to use Jennings' public toilets. There was a penny charge for using the exhibition toilets, hence the phrase "spending a penny" was introduced to the Victorian vocabulary. A penny was dropped into a door mounted slot machine to unlock the toilet compartment, providing complete privacy for the user. Male users standing to urinate into troughs, did so without charge. In ladies' toilets all compartment doors were penny operated. Gentlemen only paid a penny when requiring individual toilet compartment privacy.

Penny in the slot door lock

The Crystal Palace was re-erected at Sydenham when the exhibition closed. Jennings continued to earn £1000 a year from visitors using the toilets installed at the new site. Based on his exhibition's success in 1851, Jennings installed and staffed London's first public toilet for gentlemen in Fleet Street, followed in 1852 by a ladies' toilet in Bedford Street, Strand. Called "halting stations", Jennings charged a penny fee for their use.

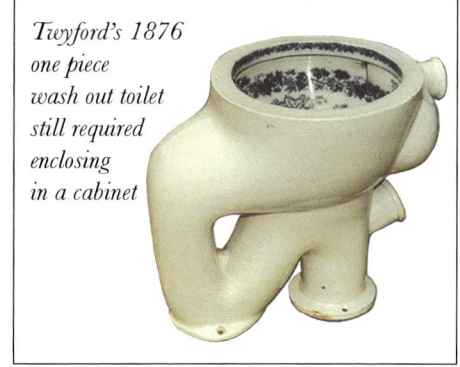

Twyford's 1876 one piece wash out toilet still required enclosing in a cabinet

Thomas Twyford was an innovative, practical potter and sanitary engineer, who set himself the objective of developing a free standing pedestal toilet using Jennings' wash out principle. Twyford and his son, Thomas William Twyford (1849-1921), worked together to create a toilet with an integral bowl and trap contained inside a supporting pedestal. By the time of his early death in 1872, Thomas Twyford had helped to make substantial progress in the one piece pedestal toilet development.

Thomas William Twyford, then aged twenty three, replaced his father in running the company. Young Twyford equalled his late father's capabilities as a ceramic sanitary engineer, in overcoming the problems of making and firing one piece toilets. The flushing action of Jennings' toilet directed water from the bowl's front, washing the tray's contents backwards over a weir into the trap. To harness the water's full power, Twyford reversed the flushing direction and outlet configuration of Jennings' wash out bowl to effectively flush forwards over the tray, clearing waste matter from toilet to the drain.

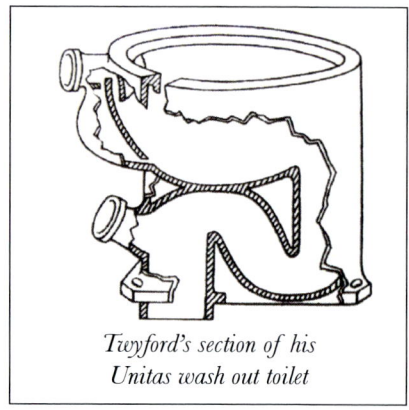

Twyford's section of his Unitas wash out toilet

The first development stage towards a one piece toilet was attaching the bowl and trap during production and successfully firing the complex piece. Early attempts left difficult to clean voids between the bowl and trap. The final stage was to conceal the trap formation within an integral pedestal. In 1883, Twyford achieved the technical and firing problems associated with this task and created the world's first one piece pedestal toilet to be made, the Unitas. Despite this advance, Twyford felt he had to provide incentive decoration to popularise free-standing toilets in Victorian bathrooms.

The progressive development stages of Twyford's Unitas wash out toilet from a two piece to an integral trap one piece until finally, in 1883, the bowl and trap are concealed inside a free standing pedestal toilet

Early water closet bowls incorporated copper spreader plates to form a film of bowl side cleansing water. The spreader cleared the tray whilst directing water around the rim forming a front cascade to clear the trap's contents. Twyford perfected the wash out toilet's flushing efficiency by incorporating an integral pottery spreader to replace the copper plate used by his competitors. Twyford developed, perfected and patented an integral pottery after flush chamber. Positioned behind the bowl, the after flush chamber retained a water reservoir which slowly drained after completion of the flush to refill the tray and reseal the trap.

Twyford's introduction of the one piece pedestal Unitas toilet in 1883 proved to be both the first and most successful free standing wash out toilet. Twyford patented his prime features of the Unitas toilet's success, the integral pottery flushing rim spreader and after flush chamber. The aesthetic appearance plus these patented attributes, gave Twyford a worldwide operational and marketing advantage over his competitors. The Unitas patented pottery flushing spreader, illustrated right, is decorated to match the bowl. The inferior white painted metal spreader on Doulton's Lambeth toilet can be seen rusting in the illustration below.

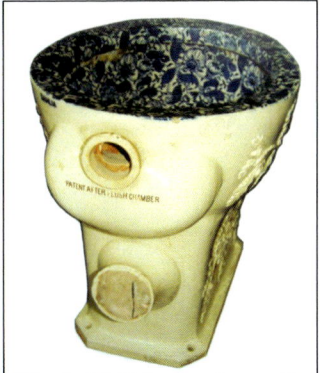

Twyfords' Unitas wash out toilet with patented after flush chamber

The case for exposed toilets was helped further when in 1886, Queen Victoria used a Unitas pedestal toilet at Doncaster's Angel Hotel. Victoria's husband, Prince Albert, had died from typhoid in 1861, and later their son almost died from the same disease in October 1871. The effect of typhoid fever on the royal household generated national appreciation of the vital role toilets played in achieving good sanitation.

The medical profession and health authorities highly commended the Unitas when displayed at the London Health Exhibition in 1886. Royal approval was endorsed with the installation of Unitas toilets at Buckingham Palace. Free standing pedestal toilets became increasingly popular in Victorian bathrooms. Queen Victoria's patronage of the Unitas toilet had a greater effect than the first Queen Elizabeth's selection of an Ajax closet for Richmond Palace.

Pedestal toilets became accepted in free standing form and gained a major market share to the exclusion of metal valve closets. The complexity of one-piece toilet production differed in concept and technology from domestic pottery. Larger toilets required new potting skills, materials and firing techniques leading to manufacturing specialisation.

Doulton Lambeth flush out toilet

By the late nineteenth and early twentieth centuries, potters such as Doulton, Howson, Edward Johns, Alfred Johnson, Johnson Brothers, Shanks and Twyford all had factories dedicated to the production of ceramic sanitary appliances. During the twentieth century many other British and overseas potters began production of bathroom sanitaryware.

Royal patronage had stimulated sales of pedestal toilets, breaking the impasse of cumbersome metal valve closets. Within two years of its launch, the Unitas outsold competitors in volume, reaching over 10,000 toilets a year. Before the nineteenth century's end, Thomas William Twyford was granted a "Royal Warrant of Appointment as Bathroom and Washroom Manufacturer to Her Majesty Queen Victoria's Government."

Twyfords' display at the Glasgow International Exhibition 1901

The twentieth century opened with the Unitas having a reputation as the industry's worldwide best selling wash out toilet. The wash out toilet became a standard type used in continental Europe and in South America.

The growing expansion of toilet sales established the generic name change from water closet or WC to toilet, considered a more up market term. In 1909, a joint Spanish stock company formed under the name "Unitas" claim they "...celebrated 100 years as having the same name as a toilet." Perhaps of greater interest, the company claimed the word "Unitas" means union in Spanish. Thomas Twyford may have been aware of this Spanish language connection when achieving the "union" of bowl, trap and pedestal to create his Unitas toilet. To gauge an indication of Twyfords' export market penetration, simply type Unitaz into a search engine to see the Russian's popular term for toilet, now pronounced "tooalyet."

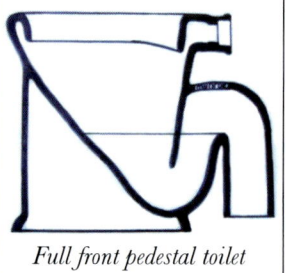

Full front pedestal toilet

Wash down toilets

In 1884, Frederick Humperson introduced the wash down toilet, considered to be the first of its type. D.T. Bostel, of Brighton, also claimed a similar invention in 1889. Wash down closets were an evolutionary development of earlier two piece Cottage pan and trap toilets, but in one piece. Early wash down toilets were full front pedestal types, but later models were modified to ease production with cut away styling. Since 1890, the wash down principle has been adopted as the prototype for most toilet development.

Cut away pedestal toilet

Inventors Humperson and Bostel co-operated with sanitary potters and no doubt all parties contributed to the wash down closet's development. Successful operation of wash down toilets depends on the force of flushing water to drive soil from the bowl. British toilets were exported around the world and acclaimed for their flushing efficiency and quality, with patterns designed to suit local customs and religious requirements.

Ceramic materials used for toilet production

Early ceramic toilet bowls were normally made from the same constituents used for domestic earthenware utensils, but for added strength they were up to three times thicker. Fired earthenware bodies were a creamy white colour and slightly porous. When dipped into liquid glaze and refired, earthenware acquired an impervious surface. The easy-to-clean glazed pottery toilet bowl surfaces were far superior to cast iron bowls that rusted through their painted finish. In addition to being more hygienic, the white ceramic colour was excellent for decoration and visually an improvement over the painted cast iron bowls they had replaced.

Stoneware

Stoneware was a cheaper, waterproof and stronger production body than earthenware. These factors convinced Henry Doulton of the material's viability. Stoneware's advantages were an almost impervious body throughout the trappage and all surfaces created in one firing; earthenware needed two. The disadvantage was that stoneware's brownish body showed through the transparent glaze and proved unpopular with Victorian householders.

Developed by John Dwight, stoneware was stable for production of large pieces. With traps used to support bowls, toilet production became more complex. These larger items required thicker and stronger construction to resist distortion during firing. Henry Doulton replaced the cast iron discharge chamber of his valve closet with a strong ceramic stoneware drainpipe body. Doulton originally bought earthenware bowls in white for his metal valve closet combinations from Pinder Bourne and Co. Burslem, Staffordshire. Doulton knew the market was moving to free standing designs and believed his impervious stoneware to be the ideal material for toilets.

These two stoneware toilets were fired with a white earthenware layer in the bowl; and up to the rim in one and rolled over the rim top in the other.

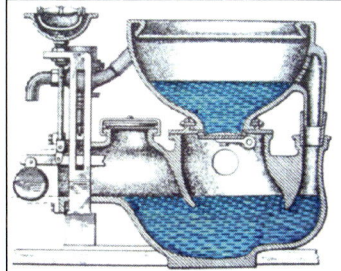

Doulton's Lambeth valve closet with a white earthenware bowl and stoneware trappage unit

Cut away toilet with incised decoration below rim

1898 full front stoneware toilet

The outside colour had not concerned Victorian customers when purchasing valve closets because 90 per cent of the unit was concealed in a cabinet, with only the bowl visible. Hoping free standing decorated stoneware toilets would be acceptable to Victorian taste, Doulton added relief and hand coloured enamel decoration. Stoneware's varying biscuit colour was associated with drainpipes and proved an unpopular alternative to mahogany cabinet bathroom décor, resulting in poor sales.

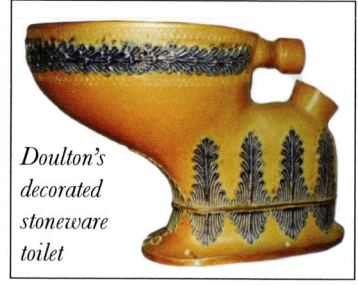

Doulton's decorated stoneware toilet

Rather than buying earthenware bowls from Pinder Bourne, Doulton stopped making stoneware toilets in Lambeth and purchased the established Whieldon earthenware pottery production unit in Stoke. This acquisition gave Doulton a presence with a dedicated sanitaryware factory in the Staffordshire Potteries.

Cane and white ware

To overcome stoneware's colour problem and retain the benefits of its strong impervious body, the buff coloured bowl's inner surface and rim were coated with liquid white earthenware. This white surface was applied to the stoneware mixture body in the clay state and fused onto it in the first firing. After firing the ware it was dipped into clear liquid glaze and refired to become cane and white ware.

The Cottage pan above shows how widely the cane and white ware colour varied and the need for it to be concealed in cabinets. The experience gained on developing effective flushing for

Variations between cane and white body

Cottage toilets helped potters in perfecting the efficiency of later wash down toilets. Cane and white became the main British production material for the volume market of working class housing toilets in the mid nineteenth century.

Earthenware

The increasing preference for white bodied earthenware gradually replaced cane ware for domestic bathrooms. Earthenware is a porous body but following the first firing it is immersed into liquid glaze and refired. After refiring the glaze is fused onto the body to form a high gloss water proof finish on all exposed surfaces and throughout the trappage. In 1886 Twyfords Limited catalogued Royal Porcelain, a type of earthenware. The phasing out of twice fired earthenware began in the 1930s with the introduction of an impervious twice fired white vitreous china body.

Twyfords' Royal Porcelain Deluge

Vitreous china

Vitreous china had similar body constituents to earthenware, but with an increased amount of feldspathic fusing material. The fusing process increased the body vitrification in the firing kiln to form a completely dense, non-porous material, ideal for sanitaryware. During the 1950s, British and American makers adopted once fired vitreous china to replace most other materials for sanitaryware production.

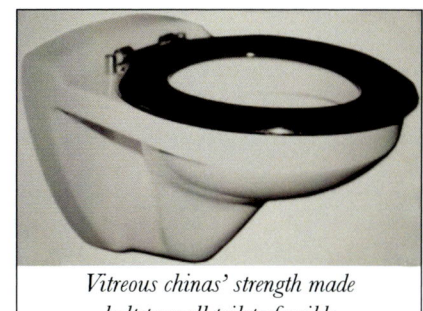

Vitreous chinas' strength made bolt to wall toilets feasible

Fireclay

Fireclay was abundantly available in areas associated with coal mining, in particular Scotland, Leeds and the Potteries. Fireclay was primarily a heavy refractory material used for kiln furniture and large heavy duty items of institutional sanitaryware. Fine fireclay sanitaryware designs were made by: Adamsez, Leeds Fireclay, Shaw Knight, Shanks Barrhead, Johnsons Fireclay, Doulton and Twyfords, but the manufacturing process was expensive and the ware heavy, therefore it was seldom considered as a material for domestic use.

Decoration

Many combinations of ceramic decoration were applied to the new free standing toilets. A selection from Gladstone Pottery Museum's toilet collection, shown below, illustrates how toilets were often transfer-decorated to match the washbasins and bidets, already established in Victorian bathrooms.

Junction and Rapidus white glazed earthenware toilets were twice fired, before applying transfers and the pieces were then fired for a third time - a fourth firing was required for any enamel painting

The lack of success in breaking into the emerging Victorian bathroom market with decorated stoneware toilets ensured that makers concentrated on white or cream earthenware for sanitaryware manufacture. Development of earthenware one piece pedestal toilets provided shapes well suited to exterior and interior bowl decoration.

Builders' merchants in the United Kingdom and agents overseas became important links between sanitaryware manufacturers and local installers. Ware was made exclusively badged with customers' or agents' names. This elegant floral decorated Junction toilet was made by Shanks of Barrhead for Giddings and Dacre, a Manchester builders' merchant.

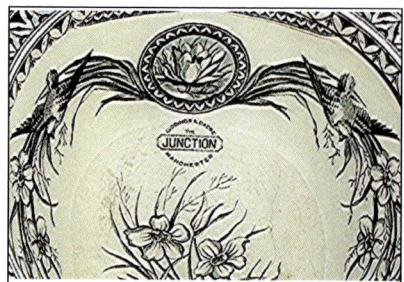

Top view of the Junction toilet showing Giddings and Dacre back stamp

The badge on the left shows the Paisley connection on the 1912 Doulton Art Nouveau toilet bowl design shown top left of the six embossed toilets illustrated on page 18

Twyfords made and badged a large proportion of ware for merchants and agents around the world. Badges reveal interesting facts about makers, for example, in addition to Lambeth and Burslem, Doulton also had a manufacturing presence in Paisley, Scotland.

Before potters' employed their own skilled labour to apply transfers, they offered raised decoration which they could cast into the ware during production. Whichever art style was currently in vogue, potters' skilled modelling expertise enabled them to reproduce classical Grecian, Roman, Rococo, Art Nouveau or simple floral décor.

Embossed decoration in the clay also guided bespoke hand painted colouring to suit individual customers' personal requirements. It should be remembered that only the very rich could afford these luxuries. Potters had at last secured a dominant share of the luxury bathroom market.

Gladstone Pottery Museum's priceless examples of a Deluge toilet and toilet paper box are decorated with hand enamelled Chromotone Pâte Dure in blue and gold tracing, demonstrating Potteries craftsmanship

Embossed decoration was used to provide interest in the less expensive market sectors with the intention of promoting free standing toilets. These examples show the variation of raised decorative designs available. The motifs were modelled into the negative plaster of Paris moulds and cast into the clay ware during the production process.

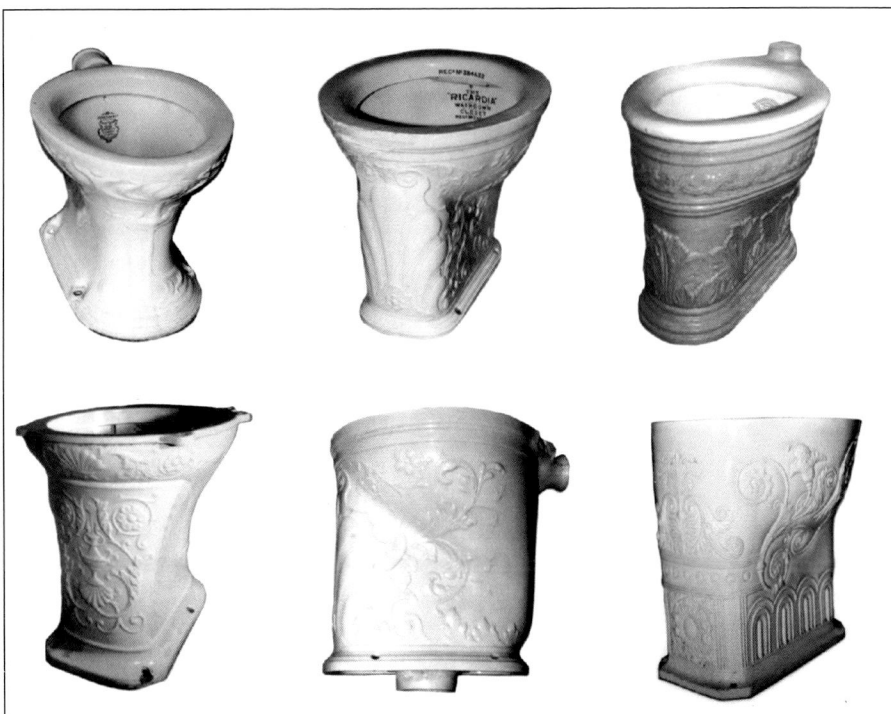

Six examples of embossed decoration cast into the clay ware during the production process; illustrating examples from across the industry, including Doulton 1912 Art Nouveau, Johnson Brothers Ricardia wash down, Shanks cane and white toilet, four seat hole Darrar and Twyfords Deluge

Embossed decoration was also used as a guide for the more expensive hand enamelling process, known as Chromotone Pâte Dure. A wealth of operatives with skills in hand painting and transfer decoration of pottery was readily available in Staffordshire. Victorian catalogues illustrate most toilets, washbasins and bidets were available with a wide choice of decoration.

Early nineteenth century ceramic valve closet bowls had been popular with transfer decoration. Engraving copper plates for making transfer prints was a well developed expertise in the Staffordshire Potteries table ware industry. Ceramic pedestal toilets provided an ideal canvas for adding the same decorative process onto their visible surfaces. Matching transfer decoration on other bathroom sanitaryware helped to spread the expensive engraving cost.

The raised motifs of embossed decoration was often used with the addition of coloured transfers to satisfy the Victorian constantly changing artistic requirements. Transfer decoration was not the only treatment of embossed motives, coloured Chromotone Pâte Dure decoration was the most luxurious form. Priceless examples of this skilful Potteries hand painted enamelling and gold tracing craftsmanship is displayed at the Gladstone Pottery Museum.

The examples above have embossed designs with transfers fired over them; this process required a total of at least three firings. The embossed motifs on Twyfords Twycliffe toilet below left were formed during casting. After a first firing, the toilet was immersed in liquid glaze and refired to fuse the glaze onto the white biscuit ware. The resultant white gloss surface formed an impervious finish on all surfaces and throughout the trappage. This smooth white glazed finish was ideal to accept the three colour hand enamelling before a final firing.

Prior to 1911 there was no provision of seat fixing holes in the ware.
Seats were either supported on cast iron or wooden wall or floor mounted brackets

The blue transfer decorated toilet below had two firings before transfers were applied and the toilet refired

This piece had the addition of a hand painted blue foot which required an additional firing

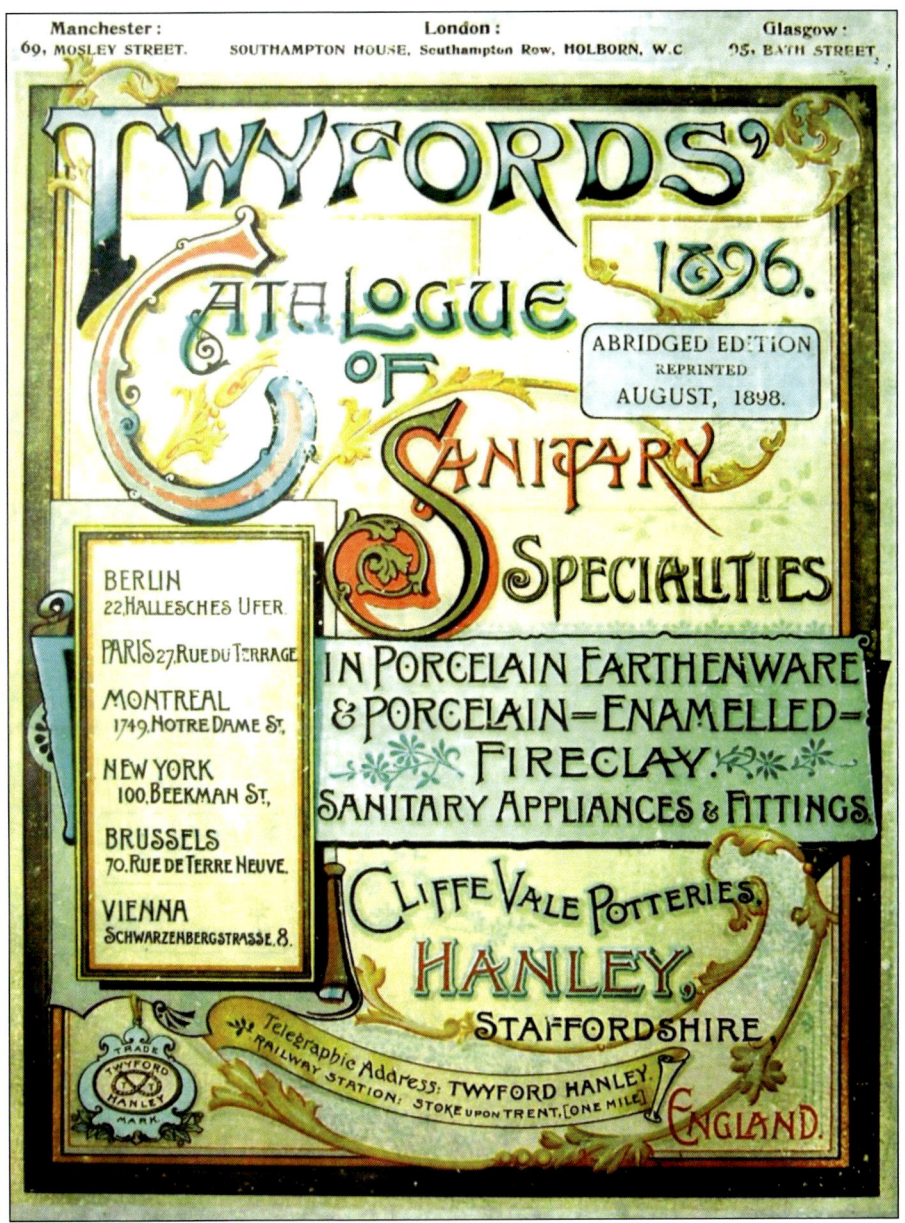

The lead page of Twyfords' twentieth century catalogue heralding the last of Queen Victoria's influence

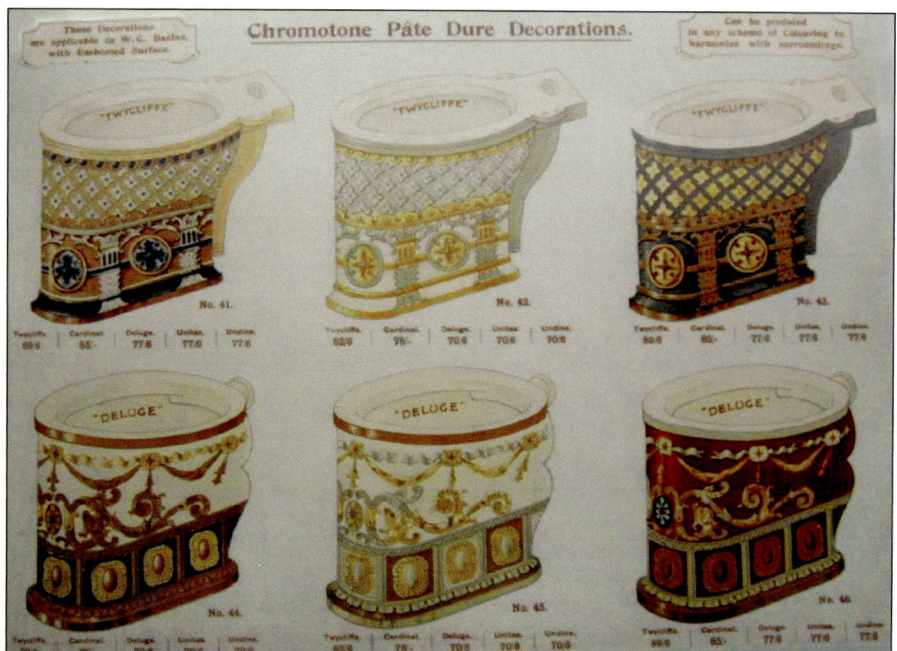

Twyford's 1898 catalogue illustrated Chromotone Pâte Dure bespoke enamelled decoration to satisfy individual customers requirements
The coloured enamels were painted over raised decoration, which served as guides for skilful hand painting craftsmanship

During the early years of the twentieth century, ornate and colourful decoration, popular in Queen Victoria's reign, began to fade. Shanks Citizen on the left and Twyfords Twycliffe on the right were hand enamelled over the embossed designs. Transfers and enamelling decoration had embraced the many genres of artistic work that became fashionable from Classical, Rococo, through Art Nouveau's sinuous plant forms, to the Modernist streamlining of Art Deco.

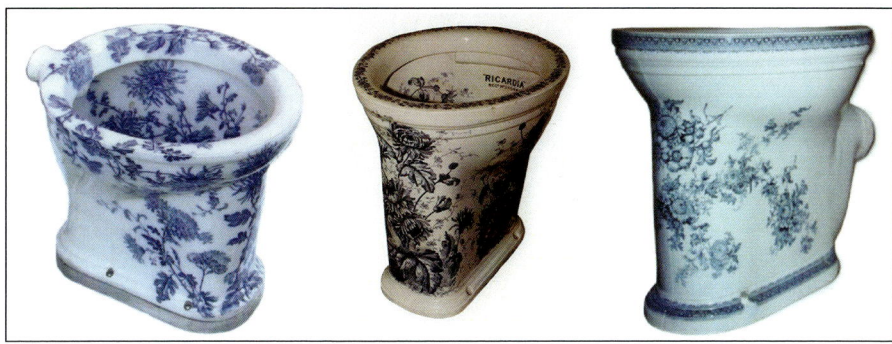

After the opening of the twentieth century potters swiftly adapted styles to satisfy the ever changing fashion of artistic tastes. Highly colourful Victorian transfer and painted ornamentation were rejected in favour of plain white. Britain's service personnel returning from the 1914-1918 Great War were promised "Homes for Heroes". These homes included white bathrooms that set the standard expected in council houses and middle class homes. The 1920s Jazz age briefly brought black banding on white ware, but by the late 1930s most forms of applied decoration disappeared in favour of the cut corner styling of Modernism.

Single coloured ware

Art Deco's Modernist sharp lines in bathroom fixture designs, with cut corners and straight sides, replaced curved edges. Pedestal toilet basins were the item that forced manufacturers into specialisation, but large luxury washbasins became the new challenge.

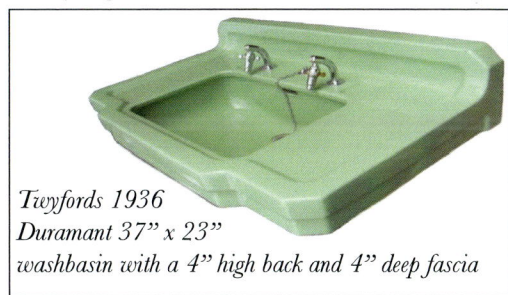

Twyfords 1936 Duramant 37" x 23" washbasin with a 4" high back and 4" deep fascia

When it was first introduced during 1925, coloured ware volume was smaller than white output, resulting in higher prices. Luxury coloured bathroom ware was priced beyond the reach of lower paid workers and some middle class buyers. Art Deco briefly popularised black and peony sanitaryware, but European and North American markets preferred the pastel colours of turquoise, ivory, lavender, primrose, blue, pink and green as shown on the cut corner washbasin above.

Shanks' angular foot

Twyfords' cut corner toilet

Flushing efficiency, syphonic and close coupled toilets

The availability of piped water in eighteenth century Britain made the filling of roof space water storage tanks possible. This provided gravity fed supplies to all floors. These water reservoirs, at constant pressure, allowed the installation of high level cisterns connected to toilet bowls by a flushpipe. High level cisterns, fitted two metres high, provided a constant supply of flushing water and replaced water supply control valves for flushing toilets. To utilise the kinetic energy of cleansing water and to contain spillage, toilets were made with full or box flushing rims. Pulling the cistern chain lifted the outlet rubber valve from its seating, allowing water to discharge into the flushpipe. Upward thrust from the float held the valve off the outlet seating until the water level dropped, lowering the valve back in place.

Grit and lime gradually built up on the outlets preventing the valves from seating correctly, the failure allowed wastage of water. A similar wastage situation had been experienced with metal valve closet outlet seals in the early days of toilet development. With the intention of keeping their toilet fresh and clean, householders occasionally abused flushing cistern valve fittings by securing the pull chain. This action left the outlet valve slightly open, allowing a constant trickle of water into the bowl and trap. This deliberate abuse led to substantial wastage of the meagre water supplies.

Twycliffe catalogue page

In the 1870s the water wastage problems were overcome by development of the valveless syphon to replace the leakage prone fulcrum valve cistern fittings. Valveless syphons were also operated by pull-and-let-go chains. The difference of the water saving feature offered by a syphon was that holding down the chain was ineffective in obtaining more water. Pulling the chain of a valveless syphon raised the volume of water contained in the syphon bell. This syphonic plug of water fully charged the syphon's down leg, drawing the cistern's contents behind it. This process started the flushing action to empty the cistern by gravity and cleanse the toilet bowl. When the water level dropped below the syphon base, air was drawn into the bell stopping the flushing action.

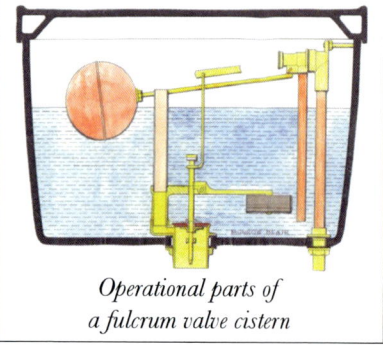

Operational parts of a fulcrum valve cistern

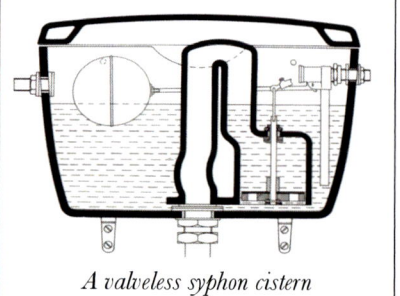

A valveless syphon cistern

Thus valveless cisterns cured a major contributor to Britain's water wastage problem. English water authorities included valveless syphons in their regulations as essential elements of water waste prevention. Syphons had no valve to leave open, so the cistern could not be operated again until it refilled.

Many plumbing engineers contributed towards the successful development of the valveless syphon, at the time termed the Water Waste Preventing syphon, known in the trade as a WWP. Doulton, Edward Johns, Shanks and Twyfords all catalogued water waste preventing syphons. Thomas Crapper was one amongst many plumbers who installed valveless syphons, but he was not the inventor. Thomas Twyford's records indicate a crate of tea being delivered each Christmas to Crapper in exchange for supplying most of their sanitary ware requirements.

Crapper's valveless waste preventer

The valveless syphon provided a constant and repeatable flushing control that helped potters to further improve toilet flushing efficiency. These gradual improvements allowed flushing cisterns to be lowered, but many remained encased in mahogany, often extended into toilet seat arrangements. Hinged wooden seats for pedestal toilets were secured to the wall or floor on cast iron brackets. Seats were not bolted to the toilets, instead they rested on the basins that in turn were floor mounted. In 1911 toilet basins were first made with lug holes for direct seat attachment.

These two examples show alternative arrangements for provision of wooden seats on pedestal toilets before 1911

The seat above is an integral part of the wooden cistern casing and supported from the floor. The seat on the right is supported on ornamental cast iron floor mounted brackets, with the seat simply resting on the toilet top.

Syphonic toilet development

The flushing actions of wash out and wash down toilets depends on water driving, or washing the bowl's contents through the trappage. To achieve effective clearance of toilet bowls, many inventors concentrated their flushing actions on syphonic principles. Syphonic toilets were designed to rely on the creation of a partial vacuum, or water build up within the trappage to pull effluent through the trap into the drain.

Profits created from his public toilets helped George Jennings to finance development work on syphonic toilets. George Jennings' syphonic Closet of the Century won a Gold Medal at the 1884 International Health Exhibition, South Kensington, London.

The syphonic action of Jennings toilet depended on most of the flushing water discharging directly into a separate second trap below the floor level, without passing through the bowl. This operation induced air displacement to pull the bowl contents into the second trap to be flushed through into the drain. George Jennings' Closet of the Century was very expensive, used a lot of water and had complicated pipe work. The unsightly syphonic action valve apparatus necessitated it being concealed in a cabinet. The water wasteful features of Jennings' closet were contrary to the potters' promotional objectives. Potters strove to develop free standing pedestal toilets that used a minimum water volume for maximum efficiency. Far more effective but less complex syphonic action toilets soon replaced Jennings' invention.

Compared with Jennings' spaghetti of pipes, *above right*, the rear view of Twyfords' Twycliffe syphonic toilet, *right*, is uncluttered and easy to wipe clean. Twyfords' twin integral flushing jets forced water through the trap towards the outlet. Fed by a single flushpipe, these water jets induced a powerful syphonic action by displacement. The combined action of drive and pull cleansed the toilet, which was designed to stand free of a wooden cabinet.

The pipework of Jennings' Closet of the Century needed concealment within a cabinet

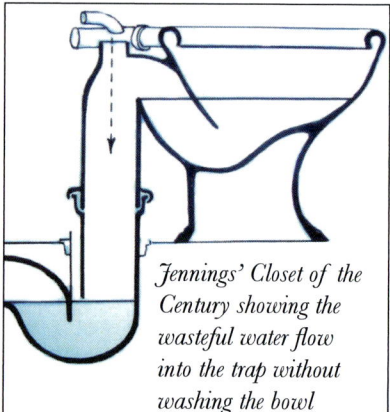

Jennings' Closet of the Century showing the wasteful water flow into the trap without washing the bowl

The syphonic action was induced by air displacement pulling the soil into and through the trappage

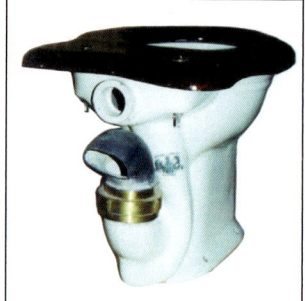

Twyfords' Twycliffe - rear view

A potter finishing a Twycliffe clay state toilet at Twyfords' Cliffe Vale factory in the 1900s

Toilet bowls with the Twycliffe type flushing action became known as "Syphon Jet" toilets. Although "Syphon Jet" type toilets are no longer made in Britain, the action remained a feature of American influenced sanitary potters.

Single trap syphonic toilets

British inventors developed other efficient methods of toilet clearance based on induced syphonic action. Harnessing wash down action principles, but instead of a full bore trappage, the out flow was retarded by trap distortion or reduced bore.

A syphonic action was created when the retarded liquid flow built up a temporary vortex within the bowl. The gravitational pull from the fully charged trappage, helped by atmospheric pressure on the bowl's large water area, forced the contents through the trappage.

These toilet types are variously known as single trap syphonic, or in the United States either reverse trap syphonic, or syphonic wash down. Variants of single trap syphonic toilets remain in production into the twentyfirst century.

This bulbous trap toilet, shown right, temporarily retards the flush through the trappage. The action allows a syphonic plug of liquid to fully charge the trap, pulling out the sump's contents.

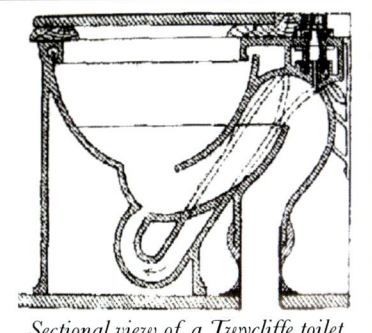

Sectional view of a Twycliffe toilet

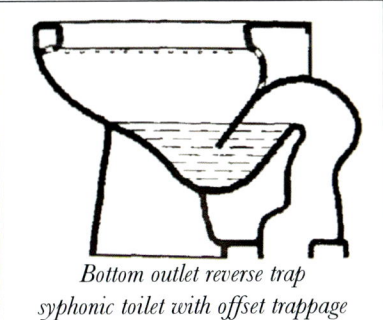

Bottom outlet reverse trap syphonic toilet with offset trappage

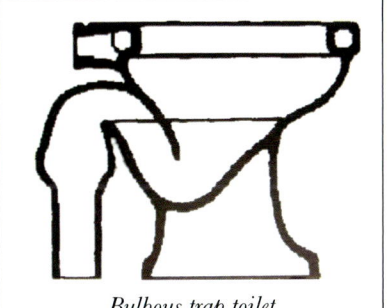

Bulbous trap toilet

Double trap syphonic toilets

By the early 1900s the introduction of syphonic closets made possible the use of low level flushing cisterns, fitted below a metre in overall height. According to a 1920s catalogue, the flushing sound of a low level double trap syphonic toilet was "practically inaudible outside the compartment". Replacing the noisy cacophony produced by metal valve closets, silence became an important selling feature together with the "extra large water area" to prevent fouling the bowl's sides. The water area is similar in size to the 1883 Unitas toilet.

Double trap syphonic toilets incorporated Jennings' two water sealed trap principle. Unlike Jennings' design, the 1930s double trap syphonic toilets reduced the internal air pressure between the traps. This utilised the whole volume of flushed water to cleanse the bowl. The syphonic effect operates throughout the flush, ejecting its contents and resealing both traps. The double trap syphonic type earned the term continuous action syphonics. Through improved ceramic casting methods, potters were enabled to include the twin traps within the toilet's pedestal.

Early double trap syphonic toilets operated under low level cisterns, delivering water through a flush pipe which incorporated an air suction pipe with an internal hooded restriction. Water flowing down the restricted flushpipe created an air pressure differential in the sealed chamber separating the traps. The resultant Venturi effect generated a powerful suction of air from between the traps. The partial vacuum drew the bowl's contents through both traps into the drainage system. The Venturi principle operating in double trap syphonic toilets is similar in action to that of a petrol engine carburettor.

Close coupled toilet suites

The expanding twentieth century North American sanitaryware industry adopted large volume valve flushing mechanisms and in some cases used mains water pressure to induce efficient evacuation. The increased volume, pressure and velocity gave American makers the advantage of bolting cisterns directly onto toilet basins as close coupled suites.

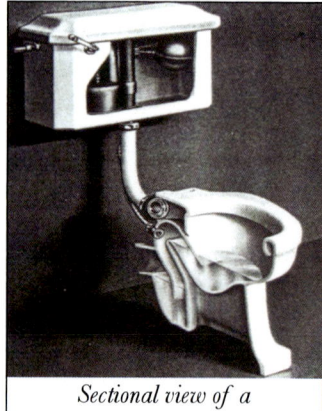

Sectional view of a double trap toilet

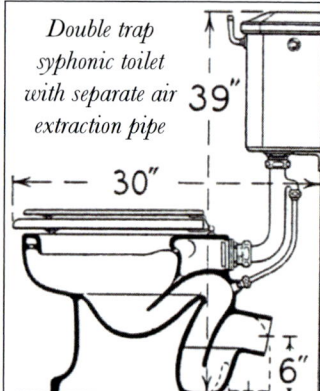

Double trap syphonic toilet with separate air extraction pipe

Bottom outlet close coupled toilet

Twentieth century British water regulations insisted toilets must operate efficiently with a maximum two gallon water flush delivered through a "Water Waste Preventing valveless flushing syphon." British toilets are not allowed to be supplied directly by mains water. Unlike their American competitors, British sanitaryware makers were forced to develop more efficient flushing from reduced flow syphons before being able to bolt cisterns to toilet basins.

This reduced capacity factor drove British manufacturers to create inventions compensating for the loss of "water head" when designing close coupled toilets. British developments brought the advantage of efficient flushing with the minimum water usage. Flushing technology and efficiency continued to improve in Britain until, by the 1930s, both wash down and syphonic toilet basins could operate efficiently, with the cisterns bolted to basins as close coupled toilet suites.

A major achievement resulting from the loss of water head was the close coupled double trap syphonic toilet. A more sophisticated Venturi pressure reduction system was required. Twyfords' Technical Designs Director, Arthur Victor Pimble, rose to the challenge by creating a totally new enclosed Venturi pressure reducing fitting in 1930.

Pimble's patented pressure reducing fitting was inserted into the base of a valveless syphon between the cistern and the toilet basin. This bullet shaped fitting created an even more efficient Venturi effect than the flush pipe pattern. Twyfords "Unitas Silent" double trap syphonic toilet suite became the British industry's bench mark for luxury design.

The protection Pimble's patent provided was a "totally enclosed pressure reducing fitting", securing a substantial market lead where no other maker could enclose their pressure system. Until the patent expired, competitors were required to have an external air pipe pressure reducer to draw air from between traps.

Twyfords Brampton double trap syphonic toilet

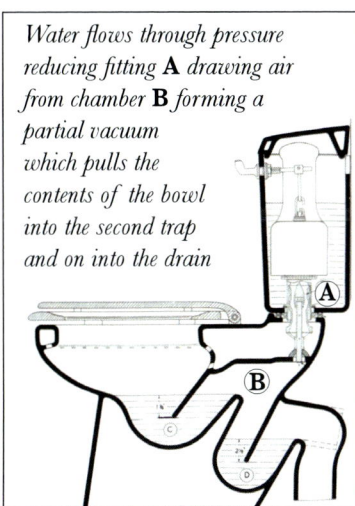

*Water flows through pressure reducing fitting **A** drawing air from chamber **B** forming a partial vacuum which pulls the contents of the bowl into the second trap and on into the drain*

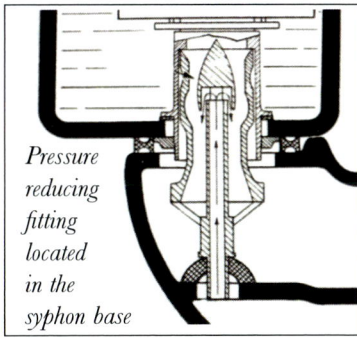

Pressure reducing fitting located in the syphon base

In 1966 "*The Lancet*" medical journal praised the Twyfords Unitas Silent double trap syphonic toilet suites for their negligible dispersal of bacterial aerosol when flushing. This health feature satisfied the growing awareness of hygiene and safety in the home. Double trap syphonic toilets have the unique efficiency of economic use of water, quietness in operation and a hygienic action to prevent splashing or dispersal of airborne bacteria.

American bathroom manufacturers

North America became a lucrative market for British sanitaryware makers from the middle of the nineteenth century. America's first pan closet was made in the 1830s. Production expertise was probably influenced by immigrant English craftsmen from the Staffordshire Potteries. The United States buying public considered early American domestic production inferior to English sanitaryware, despite the high cost of importing. Thomas Twyford's ware was revered by Americans as the most efficient and best quality English bathroom fixtures.

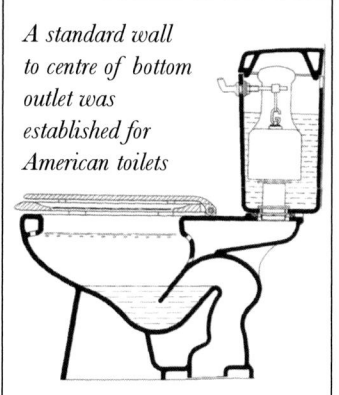

A standard wall to centre of bottom outlet was established for American toilets

The White House had its first bathroom in 1850. Forty three British potters were exporting high quality sanitaryware to the burgeoning United States market in 1873. Early American sanitary potters continued to find it a hard job to secure a home market share. Zane and Co, Boston, introduced a sanitary closet bowl with galvanised plunger chamber in 1881. A. G. Myers, New York, patented a one piece earthenware solid plunger closet, but these early American toilets lost prominence to Thomas Twyford's one piece Unitas Pedestal toilet launched in 1883. By 1895 George Washington Vanderbilt's famous Biltmore House in Asheville, North Carolina, was completed with fifty seven bathrooms, each with English ceramic flushing toilets and bathroom fixtures.

American sanitaryware potters quickly reacted with improved product quality and designs. Between 1900 and 1932 there were three hundred and fifty American water closet patents. American potters appreciated the vast potential of their home bathroom market, at that time being satisfied by European potters. Many ploys were adopted to market American made sanitaryware products as equal to the British. In 1912, an East Liverpool producer registered his company name as ADAMANT Porcelain Co., this was the trade name of fireclay sanitaryware made by Twyfords.

Thomas Maddock, an American producer, marked his American made products with a similar badge used by the English sanitary engineer, George Jennings. Maddock's badge incorporated a Jennings' type pseudo English royal cypher with the rampant lion and unicorn. In addition, Maddock's deceitful badge also stated: "Best Stafford Earthenware made for the American Market".

In the 1890s, several of the current American ceramic sanitaryware companies started to emerge. Metal baths were the popular starting point for Kohler in 1904 and Eljer by 1907. Meanwhile central heating radiators and water closets had set American Standard on track to develop vitreous china plumbing fixtures in 1913. American sanitaryware sales and production grew by a phenomenal 367% between 1929 and 1954.

The English double trap syphonic principle was not adopted by American sanitaryware makers, who preferred to develop bottom outlet and air pressure evacuation. The American industry created their own standards and regulatory bodies. Reverse trap syphonic toilets became the popular flushing method with standardised wall to centre of bottom outlet, known as the "roughing in dimension".

Valve cistern flushing apparatus discharged large volumes of water delivered at high flow velocity to improve flushing. These American manufacturers' volumetric flushing aids saw the development in the 1930s of one piece toilets.

A 2011 Crane one piece bottom outlet toilet - but this type had been made in the United States since the 1930s

Worldwide sanitaryware production

The seeds of industrial scale sanitaryware production were first sown by Thomas Twyford in Bath Street, Hanley, Staffordshire, England, in 1848. From this modest beginning, sanitaryware production has spread throughout the world. It could be argued that Thomas Twyford started the overseas production process at the beginning of the twentieth century to protect his lucrative exports throughout Europe, in particular to Germany. Thomas Twyford's trade had been restricted through the imposition of high German tariff barriers on sanitaryware.

In 1903, to regain his dwindling German business, Twyford built his own sanitary pottery factory at Ratingen, near Dusseldorf, to satisfy the German market and European distribution. Other British manufacturers followed suit, Johnsons of Staffordshire had a factory at Florsheim and Alfred Johnson had one at Wesel. In 1914, during the Great War, the German government sequestered the Ratingen, Florsheim and Wesel factories as enemy property, without compensation. Adding insult to injury, in 1917 the British factories were formed into the Keramag Group to become post war competitors.

In 1928, American Standard opened a vitreous china sanitaryware plant in Neuss near Dusseldorf, Germany. Ten years later, American Standard also started production of vitreous china sanitaryware at Hull, England. After the Second World War, to protect emerging local production, or through lack of exchange currency, high import tariffs were placed on imports into many traditional British export sanitaryware markets. Rather than lose the trade, several major United Kingdom sanitary potters built factories in overseas markets such as India, South Africa, Australia and Malaysia.

Following their initial slow start, North American sanitaryware producers secured the United States market, and began expanding production and expertise to plants throughout the world. In the 1950s, American influence spread worldwide, helped by the tourist hotel groups building worldwide, plus international acquisitions and joint ventures with overseas sanitaryware makers. Towards the end of the twentieth century, volume sanitaryware production had moved around the world to wherever production wages, overheads and exchange rates were low.

Development in recent years

The marketing expertise of early British sanitary potters broke the mould of tradition by establishing toilets outside cabinets, where they proudly remain free standing today. What fashion trends, design styles and technical innovations can we expect in the third millennium?

One could claim the Victorian operational methods developed so long ago need no improvement. Even the electronic cleansing facilities provided with the Clos-O-Mat combined toilet and bidet unit depends on the wash down principle to clear soil.

Gradual developments of toilet efficiency paved the way for European legislation to reduce cistern flushing capacities from nine to six litres, together with dual flush options. United States makers now also offer dual flush economy, a long stride forward from their earlier six gallon flushing.

During the 1980s American development of Pressure Assisted Flush systems began. Manufacturers catalogues explain how these toilets operate. "Water is forced out with the pressure of compressed air in the tank when flushed. The toilet uses the water pressure provided by your water company; no pumps or other devices are necessary." The pressure assisted operation works on the principle that the elasticity of air allows it to be compressed; water will not compress, if squeezed it squirts.

Foul water has to be prevented from back flowing into mains water supplies. American States that allow the use of pressure assisted flush toilets insist on precautions to avoid back flow into the mains supply. To protect water purity, this American pressure assisted system is not approved for British or most European homes.

Clos-O-Mat toilet

Toilets of the third millennium

Nostalgic Victorian style bathrooms enjoyed a twenty year span of popularity into the first decade of the twenty first century. By the dawn of the twenty first century many new producing nations began exporting to traditional British markets. Competitive sanitary pottery production moved to North Africa and the far east. Lower wages in other markets have seen the Indian, Chinese and South American producers improve their quality and increased output volume. European and North American producers were unable to compete against these low production costs, thus becoming importers from the countries once considered their customers.

Selection of twentieth century toilets showing the more bulky design trend
All toilets have valve cistern fittings to provide push button operation
The lower illustration of a wall hung toilet to facilitate cleaning under the bowl
The bottom right toilet is made in one piece, a far cry from Twyford's Unitas

Modern toilet quality, efficient operation and plumbing are far superior than Victorian standards. On the technical side, from 2001 onwards, European Community ruling allowed continental type toilet valve flushing mechanisms to be re-introduced to Britain. This brought the opportunity of button operated flush rather than a lever. Victorian prohibition of valve type cistern flushing mechanisms was on the grounds that they were prone to failure through leakage, resulting in wasted water.

Time alone will prove whether Victorian leakage experience, caused by lime deposit on cistern outlet valve seatings, have been cured on the new valves. Britain's century old water conservation measures could be jeopardised by leakage through continental type toilet cistern outlet valves. The useful days of Valveless Water Waste Preventer syphons and Britain's water conservation precautions may be numbered.

In 1999 Twyfords Limited was granted The Royal Warrant of Appointment as Manufacturers of Bathroom and Wash room fittings to Her Majesty Queen Elizabeth II. This award celebrated the one hundred and fiftieth anniversary of the birthday of Thomas William Twyford; the founder of the world's sanitaryware industry. Innovative Victorian sanitary engineers gave a good start to the development of the industry. Surely twenty first century technology can devise waste free, environmentally efficient ways to clear human excrement from our homes.

Mr Twyford's 1888 catalogue page illustrating the celebrated world's first one piece ceramic toilet
This invention united the bowl and trap within an integral pedestal support - superimposed onto the page is a surviving Unitas toilet which helped to break the monopoly held by the metal valve closets

Thomas William Twyford, Sanitaryware Potter 1849 - 1921

Glossary of names and terms 34

Ajax, Sir John Harington's water closet, a corruption of the Elizabethan term "a jakes" describing a "privy".

Ashpit, An early form of toilet without water flush, soil was covered in ash from the fire and cleared manually.

Ballvalve or Ballcock (American term), A valve to control the supply of water to a cistern.

Bathroom fixtures or appliances, ceramic ware with installed sertvices.

Bourdalou, A urinary receptacle for females. Also called coach or carriage pots. Name attributed to Pere Louis Bourdaloue (1632-1704.), a popular Jesuit preacher who served in Louis XIV's court.

Box, full or return flushing rims, on WCs were first returned, then turned down into the bowl, then boxed in.

Bramah, Joseph, Patented the second water closet, this with a hinged outlet valve.

Cane and White, Toilets made from a blend of ball and china clays with Cornish Stone inside and rim coated with white earthenware and fired, then re-fired with a clear glaze.

Cesspit, Normally a hole in the ground or cellar below a dwelling to collect effluent for manual clearance.

Chromtone Pâte Dure, Form of enamelling over embossed designs.

Cistern flushing valve, Continental toilet cistern mechanism, liable to water wastage.

Closet, As in water closet. To "close it" from the "stink trap", to block cess pit or drain odours.

Close coupled WC suite, with a cistern bolted directly to the basin.

Close stool water closet, An early type of closet made by Wedgwood.

Clos-O-Mat, Electronically operated automatic toilet and washing facility.

Coach or carriage pot, The English equivalent terms for a bourdalou female urinary receptacle.

Commode, A piece of bedroom furniture housing a jerry, to collect excretory effluent during the night.

Continuous action syphonic, The double trap pattern pulls the bowl's contents throughout the flush.

Cottage pan, A type of simple pan water closet with a bowl or pan and separate trap.

Cummings, Alexander, A Bond Street, London watch maker who patented the first water closet in 1775.

Cut away front, Wash down toilet with a recessed front, illustrated on page 13.

Earthenware, A slightly porous type of ceramic material with clear or coloured impervious glaze.

Enamelled fireclay, A ceramic refractory clay suitable for large products, autopsy tables etc..

Flushing rim spreader, An integral clay, or separate metal, plate fixed under toilet rims to form a screen of water to sweep the contents of wash-out toilet trays into the trap.

Free standing, Relates to water closets designed to stand exposed in the bathroom environment.

Full front pedestal, Toilet with the trap formation concealed by an integral skirting.

Gongfermors, Teams of men who cleared effluent from house hold cess pits.

Halting Stations, The name given to public conveniences proposed by George Jennings.

Jerry or potty, Bedroom basin with a handle for urination during the night, normally kept under the bed.

Lancet, The, Official journal of the British Medical Association.

Lavatory, Strictly speaking a washbasin from the Latin lavare to wash but a term often used for a WC or toilet.

Lead safe trays, Fitted beneath metal valve closets to prevent leaking urine seeping into the floor.

Liverpool pan, A type of simple pan water closet with a bowl or pan and separate trap.

Night soil men, Victorian teams of men who took over the task of clearing effluent from house hold cesspits.

Plunger or plug closet, Early attempts to make one-piece pottery toilet, they were poor in performance.

Plumbed in, A system connected to water supply and waste services.

Privy, An ashpit or pail closet from which soil required manual emptying.

Reverse trap syphonic, A bottom outlet toilet with a restricted trap to induce a syphonic action.

Sanitaryware, Bathroom ceramics with installed services.

Single trap, or **Wash down syphonic,** Toilet with a restricted trap to induce a syphonic action.

Slider valve, Alexander Cummings' sliding outlet plate featured in his metal valve closet patent.

Soap and sponge dishes, Dressing table top accessories normally silver or pottery.

Spending a penny, The amount charged by Jennings for use of his toilets.

Spigot, This term referred to integral hollow pottery tubes for WC supply and overflow connections.

Stink trap, Patented by John Gaittait in 1782, to prevent cess pit gases entering dwellings.

Stoneware, A dense impervious pottery body developed by John Dwight (c1633–1703) in his Fulham Pottery.

Sump, The lowest part of a water sealed toilet trap.

Syphon jet toilet, with a proportion of flushing water directed into the trappage creating a jet stream to pull soil from the sump. Known in some markets as a blow out closet.

Toilet, Name for a water closet. In Victorian times, referred to dressing table "toilet sets".

Toilet sets, Comprising soap and sponge dishes, water jugs, bowls and assorted jerry pots.

U bend, A U shaped form, filled with water in toilets to create a seal against cess pit gas.

Unitas, Name of the first one piece pedestal wash out toilet by Thomas Twyford.

Unitas silent syphonic, Twyfords double trap syphonic suite with close coupled cistern and basin.

UNITAZ, Russian Language name for water closet taken from Twyfords Unitas wash out toilet.

Valve closet, Early water closets with metal outlet valve operation.

Valveless flushing syphon, Operates by lifting water over a syphon invert to pull the contents from a cistern.

Valveless syphon waste preventer, Syphon to replace valve outlet flushing mechanisms.

Venturi, a device using fluid flow over a constricted orifice causing suction of air to create pressure differential, as in a carburettor air inlet.

Vitreous china, A dense, impervious white pottery for sanitaryware manufacture

Vortex, A whirling mass of liquid, as in the spiralling movement of water around a whirlpool or bath outlet.

Wash down syphonic, A toilet basin with a single trap induced syphonic action.

Wash down toilet, A basin with controlled flushing to meet at the bowl front and direct the flow onto the water area, driving soil through the full bore water sealed trap.

WC, water closet, lavatory or toilet, Terms for a defecation bowl cleansed by water flow. Toilet developments included toilet paper boxes in the 1880s. In 1911 toilets were made with holes for seat hinges.

Water cock, The early term for water supply control valves or taps.

Water sealed trap, Water filled U shaped formation stopping cesspit gas escaping.

Water Waste Preventer (WPP), A device to lift a body of water into free fall to start a syphonic flushing action to empty the cistern. Valveless syphons overcame water wastage and replaced leaky valve flushing systems.

37   *Places to visit*

To ensure the items you want to see are on display, it is recommended that interested visitors telephone before travelling.

Abbey Mills Pumping Station, Corporation Street, Leicester

The Crossness Engines Trust, 43 Castlefield Ave, London, SE9 2AH. (Visits by appointment)

Gladstone Pottery Museum, Longton, Stoke-on-Trent. The Ceramic Sanitaryware Gallery of this working museum houses one of the best displays of historic water closets in Britain. A selection from the Gladstone collection are illustrated below Telephone: +44 (0)1782 237777

Manchester Museum of Science and Industry, Liverpool Road Station, Castlefield, Manchester M3 4JP

Museum of the History of Science, Old Ashmolian Buildings, Broad Street, Oxford, OX1 3AZ

Newarke Houses Museum, The Newarke, Leicester,

Technology Group, Science Museum, Exhibition Road, London SW7 2DD, (A selection of toilets, valveless syphons and associated drain pipes).

Collectable toilets